A NOTE FROM DR PHIL TRATHAN
Emeritus Head of Conservation Biology, British Antarctic Survey

Earth observation satellites allow scientists to see our planet from space, including our cities, fields, forests, deserts and oceans. These satellites take images in a similar way to that of a camera taking a photograph. Modern satellites reveal the Earth in such detail that things as small as the book you are reading are visible. Scientists use satellite images for many reasons, including looking for wildlife, such as Emperor penguins. Emperor penguins breed in the Antarctic, and before satellites we knew of only 36 breeding sites. Now we know of 61 sites, mostly on the frozen ocean: sea ice. Emperors are clearly visible in satellite images because they stand out against the white background of ice and snow. Scientists believe there are just over 256,500 pairs of Emperor penguins.

When they are breeding, Emperors face a number of dangers, including unusually fierce storms breaking up the sea ice, like in this story. If the chicks only have downy feathers that are not waterproof, they can be lost in the icy cold ocean. Scientists also now believe that Emperor penguins are at risk because humans are changing our planet in a way that causes warming of the air and the ocean. Warmer air and warmer oceans cause sea ice to melt, including where Emperor penguins breed. Over the next 80 years, as our planet warms and sea ice disappears, risks to Emperor penguins will become very much greater. Satellites are helping us tell this story.

For Cordie, Alex and Edward Lemon and Hudson and Sistene Trathan with best wishes, and with thanks to Dr Phil Trathan for enduring many questions. **N.D.**

For Daniel and William. **C.R.**

First published 2023 by Walker Books Ltd, 87 Vauxhall Walk, London SE11 5HJ • 10 9 8 7 6 5 4 3 2 1

Text © 2023 Nicola Davies • Illustrations © 2023 Catherine Rayner • The right of Nicola Davies and Catherine Rayner to be identified as author and illustrator respectively of this work has been asserted in accordance with the Copyright, Designs and Patents Act 1988 • This book has been typeset in Filosofia and Avenir • Printed in China

EMPEROR OF THE ICE

NICOLA DAVIES

illustrated by
CATHERINE RAYNER

WALKER BOOKS
AND SUBSIDIARIES
LONDON · BOSTON · SYDNEY · AUCKLAND

It's April in Antarctica. This isn't spring, but autumn – the sun sinks lower every day, and soon it will be gone for three whole months. Most animals are heading north to avoid the coming winter, and the ring of frozen ocean that grows with every passing day.

But that sea ice, locked tight onto the coast by the bitter cold, is what the Emperor penguins have been waiting for.

In places so remote that they can only be seen by satellites in space, they start to appear …
lines of penguins and their shadows, like spidery writing across a blank page.

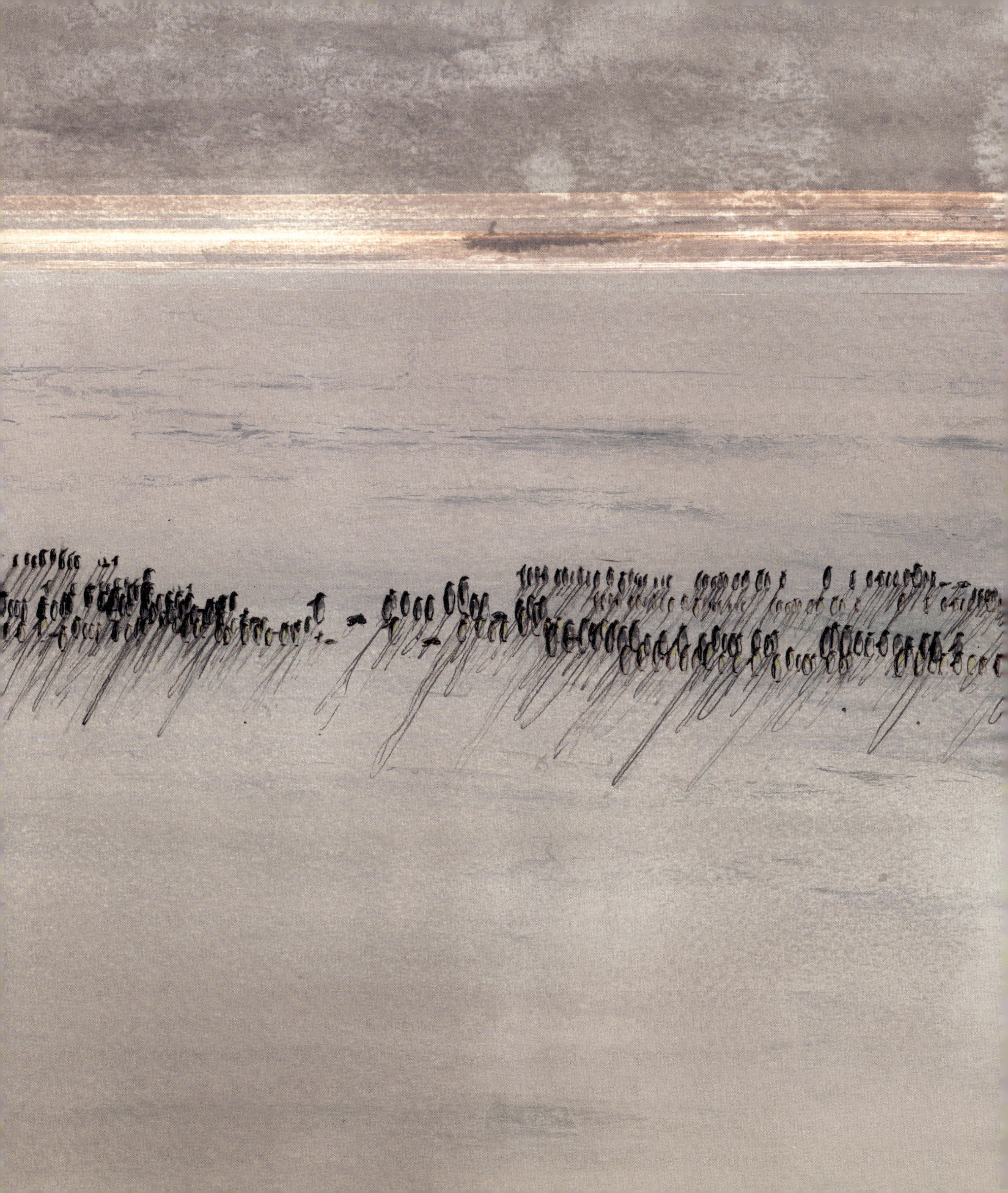

They've come to have their babies, and they need the sea ice to breed.

Not just any ice will do: it must be low enough to leap onto from the ocean; it must be smooth enough to walk over, and slide on; and it must last through winter into early summer – just long enough to raise a chick!

The ice at Halley Bay is perfect... It's ice a penguin can rely on.

That's why this wise old Empress comes back here, year after year.

But there's no time to waste, because she needs to find a mate!

And there are twenty thousand males to choose from, all trilling,

purring, singing. She listens hard to choose a voice she likes.

They mirror each other's movements, touching heads and

beaks and flippers, and bowing low ... building a bond

that makes these two a team.

The days shrink down to nothing: by early May the sun has gone, and the cold
is biting hard.

   This is when the Empress lays her single egg, huge and white. There is no cosy
nest for it – and, without protection, the egg will freeze. But this is where the teamwork
starts… At once, her mate scoops it into his pouch; protected here, the chick inside
can grow.

Laying the egg took almost all the Empress's strength. She must head back to sea
to feed, and leave her mate and egg behind, or she will die.

Through ten long weeks of deepest, fiercest winter she swims through many hundreds of miles of stormy ocean, hunting fish …

while he huddles, hungry, as the blizzards rage at his back.

No human eyes are there to see it, not even a satellite can peer through the winter dark –
but the bond between the Empress and her mate holds firm. And in July, it brings her back.
She's sleek and strong again, with a belly full of fish … to feed their newly hatched chick!

All around, the colony is celebrating after weeks of quiet – pairs greet each other noisily.

But there are pools of silence… Some males have starved; some females did not make it

back. And some eggs have frozen. A reminder that Antarctica is a tough place to survive.

The newly hatched chick is just a hundredth of the Empress's weight — it's much too small to keep warm on its own. Until it's bigger, its parents must take turns: one to babysit and one to hunt.

For weeks the Empress and her mate's journeys criss-cross the ice, until at last their chick is big enough to keep warm inside its own downy coat. This means the Empress and her mate can both hunt, to keep up with their baby's growing appetite!

The sun returns. September days stretch out into spring … and the colony is busy, noisy, full of life. Fluffy chicks explore their world, and adults call them to be fed. By December they will have feathers that are waterproof, in time to leave the melting ice and find food in the ocean for themselves.

It's been a good year for this colony.

But Halley Bay is no longer a safe place for Emperors to rear their young.

Climate change has made the seas warmer, the weather fiercer.

October storms pound the sea ice into pieces. Next year the Empress will have to find a new place to raise her chick.

It's April, again, in Antarctica.

At Halley Bay, the satellite looks down onto a page that's almost blank. Thousands upon thousands of Emperors have disappeared. But wait: look down. In other places …

there and there ...

and there.

Tiny colonies of Emperors have grown huge in size, while other new ones have sprung up.

Spidery lines of penguins have scribbled into hundreds and thousands.

Somewhere down there, the Empress has found a new place to raise a chick: a place

where the sea ice can be trusted, where there's ice she can rely on.

At least, for now.

Emperor Penguins and Climate Change

Emperor penguins are the biggest penguins on Earth and are wonderfully adapted to life in their harsh Antarctic home. But climate change is making the Antarctic Ocean warmer: winds are stronger and storms fiercer. This can break up the sea ice that Emperors need to raise their chicks.

In September 2016, the ice supporting one of the largest colonies of Emperor penguins (at Halley Bay in the Weddell Sea) broke up in a storm, and many were lost. Since then, no Emperor penguins have bred at Halley Bay... And, with the world warming up, they are in danger of disappearing.

But we can still do something to help. Satellite pictures have shown that the adult penguins that escaped from Halley Bay may have found other places to breed: if human beings can protect their fishing grounds and fight climate change, Emperors can do the rest. They are survivors.

What *IS* Climate Change?

The air that surrounds you right now is part of what we call the Earth's atmosphere: a layer of gases that stretches from the ground to more than 100 km above your head, and which protects us from the cold of outer space. But human beings have burned so much coal, oil and gas to make energy to run our homes, farms, cities, cars, planes and trains, that the mixture of gases in the atmosphere has changed. Now there is too much of a gas called carbon dioxide in the air, which affects the balance and means that Earth's atmosphere has trapped too much heat. As a result, the planet is getting warmer.

THIS is climate change, and it's causing unusual weather of all kinds, which makes life more difficult for both animals and people. Storms are bigger, floods and droughts happen more often, and the ice at the North and South Poles is melting, making the sea levels rise.

What Can People Do to Help?

All around the world, people of all ages are working to make a difference about climate change. Lots of things we do in our ordinary lives use energy and add to the carbon dioxide that is in the atmosphere … but the good news is, that means there are little ways families can help – things like using less energy at home, by turning off lights when they aren't needed and not wasting food or water.

You can talk to your parents and teachers and come up with ideas together: for example, you could ask if your school gets electricity from a company that uses solar, wind or wave power, or do something to support an organization that protects the environment and helps take carbon dioxide out of the air. Most important of all is to tell other people what you know about climate change and what they can do to help!